Contents

What is Explore Geography?

Explore Geography gives you all the resources you need to teach Geography at Key Stage 2. The pupil books cover eight core study units, with a focus on the QCA Schemes of Work. Each pupil book is supported by a teacher's guide that provides a range of resources and activities.

These books are complemented by a CD-ROM or website for each study unit. The website address is www.heinemannexplore.com. These resources give a huge library of information that will stimulate geographical enquiry and enable you to incorporate ICT in the Geography curriculum.

Pupil book

The *Investigating the Weather* pupil book includes the following features:

- **Activities** to aid independent learning and research
- **See for yourself** boxes to encourage pupils to get out of the classroom and into the field
- **Exploring further boxes** give ideas for further research and provide a link to the extensive geographical resources available on the Explore Geography CD-ROMs or website
- **Maps** to help develop pupils' geographical skills.

CD-ROM and website

The Heinemann Explore Weather CD-ROM and website contain a wealth of resources. They include the following sections:

- **Exploring**: detailed text about the area of study.
- **Global issues**: text answering questions from the QCA Schemes of Work.
- **Digging deeper**: a library of text ideal for research or topic work.
- **Pictures**: photos, illustrations, and other images to help develop geographical awareness.
- **Resources**: videos, animations, maps, and interactive pictures.
- **Activities**: to extend knowledge and further explore concepts introduced in the text.

How to use this teacher's guide

This teacher's guide offers resources to complement the pupil book and CD-ROM or website for this study. It includes:

- **A curriculum reference grid**: showing how the components of *Explore Geography: Investigating the Weather* match the Key Stage 2 National Curriculum and QCA Schemes of Work.
- **Teacher's notes and worksheets**: see 'Using the teacher's notes and worksheets' below.
- **Further research**: contains books and websites that point pupils and teachers towards useful resources.

Using the teacher's notes and worksheets

There are thirteen sets of teacher's notes and photocopiable worksheets.

The teacher's notes are set out in the following sections:

- **Outcome.** This sets out what the children will be aiming to achieve during the core activity.
- **QCA Schemes of Work objectives.** This section lists the relevant objectives for each spread of the book.
- **Keywords and concepts.** This is a list of words and concepts that most children should be familiar with when the unit is finished
- **Lesson plan.** This list suggests how to start and finish a lesson relating to the activity on the worksheet.
- **Core activity.** This is a detailed explanation of the activity on the worksheet.
- **Follow-up activities.** This suggests a range of additional pupil activities linked to the topic.
- **Exploring further.** This gives the teacher signposts to relevant pages in the Explore Geography CD-ROM or website. The themes can be researched by the children independently on the computer.

Photocopiable worksheets

The photocopiable worksheets are ready to give to the children. They have all the instructions the children need, but are not designed to be given to them without discussion or introduction. At the end of the teacher's notes and worksheets there are photocopiable maps that can be used with the pupils.

	QCA unit	Geographical enquiry and skills												Knowledge and understanding of places							Knowledge and understanding of patterns and processes		Knowledge and understanding of environmental change and sustainable development	
Pupil book unit		1a	1b	1c	1d	1e	2a	2b	2c	2d	2e	2f	2g	3a	3b	3c	3d	3e	3f	3g	4a	4b	5a	5b
What is the weather?	**7.4 7.6**	PB WS	PB	TG			PB WS	PB		WS TG									TG					
How do we know what the weather will be like?	**7.6**	PB	PB WS TG	PB		WS	PB WS		WS TG	PB WS	WS	WS							TG					
How can we measure the weather at school?	**7.4**		TG	TG		TG	PB	EF PB WS TG		EF	WS	EF TG	TG	PB										
Sunny weather	**7.4**	TG	PB	PB			PB TG		EF	EF TG	WS	EF TG		TG										
Rain and snow	**7.6**		EF	EF		WS	PB WS			EF		EF TG										PB		
Windy weather	**7.4**		TG	TG		PB	PB		WS	TG			TG									PB TG		
Weather around the world	**7.2**	WS PB EF		PB			PB		PB WS TG	EF		EF		PB WS			PB		PB TG	TG				
How does the weather affect human activity?	**7.6**	PB	TG	PB WS						TG		TG									TG	PB TG		
Why to people go on holiday?	**7.6**	PB	PB WS	WS		TG		TG	WS	EF TG		EF		PB WS	PB	PB	PB EF					TG		
Where can we go on holiday?	**7.3**	PB WS	PB	WS	PB WS									WS										
What do we need to take with us?	**7.5**	PB	WS	PB WS		PB TG				PB WS			PB WS	PB WS	PB				TG	PB				
How are places similar to and different from our own locality?	**7.4**	PB WS	PB WS	WS	TG	TG			WS					TG	PB	PB	TG		PB WS					
Which places have we visited?	**7.4**	PB WS	PB WS			PB WS				WS		TG		PB WS	PB WS	PB WS								

National Curriculum and QCA Schemes of Work correlation chart

KEY

PB Pupil book

EF Exploring further box in the pupil book

WS Pupil worksheet

TG Follow-up activities in the teacher's guide

What is the weather?

(This section links with pages 4 to 5 of the pupil book.)

Outcome

During this lesson the children will think about the different types of weather and write questions around a picture of their choice that shows a type of weather. These questions can then be used for focused research about the type of weather shown on the picture.

QCA Schemes of Work objectives
Unit 7, Sections 4 and 6

Children should learn:

- about weather conditions around the world
- to use geographical vocabulary
- to ask and respond to geographical questions
- to use secondary sources
- to use ICT to access information.

Key words and concepts

weather	**atmosphere**	**temperature**	**thermometer**
water vapour	**humidity**	**weather machine**	

Lesson plan

1. Have the following resources available: photographs illustrating different types of weather.
2. Explain the lesson objectives to the children.
3. Either in pairs or as a whole class, ask the children to brainstorm what they know about the weather.
4. Discuss how they know about different aspects of the weather (e.g. observation, books, television, pictures, Internet, etc.)
5. Explain/discuss particular key vocabulary words. It would be useful to have pictures/photos to illustrate and/or correct any common misconceptions.
6. Ask the children to do the core activity (Worksheet 1).
7. Get the children to show their work and explain the reasons for their choice of weather type. Ask them to use secondary sources to help their investigation.
8. Invite the children to exchange and share their findings.

Core activity – Worksheet 1
(page 5)

In this activity the children are given copies of photographs illustrating different types of weather. They should choose one to stick on their worksheet. They will need to think about the weather shown and write questions around their photograph. These questions can then be used for focused research. The Heinemann Explore CD-ROM or website might be a good place to start when doing any research online.

Follow-up activities

- Enable the children to use the Internet to find a painting of the weather. Get them to describe the weather in the painting and give reasons for their choice.
- Ask the children to choose a picture from the Internet or a newspaper/magazine showing a particular type of weather that includes people. Get them to print it out and stick it in the middle of a big piece of paper. Ask the children to write in speech bubbles what the people might be saying about the weather and the effect it is having on their life and the local environment.
- Ask the children to begin to make a collection of stories and poems about the weather. Get them to select and sort these according to weather type. Do they notice any patterns emerging? Is any one type of weather written about more than the others? Did they find any particular type of weather with few examples?

Exploring further

The children can find out more about the weather by using the following links on the Heinemann Explore Weather CD-ROM or website:

- Exploring > Weather around the world > Weather and climate
- Digging deeper > What is weather? > Watching the weather
- Digging deeper > Weather > Have a nice day
- Digging deeper > Weblinks

Worksheet 1

What is the weather?

Name ________________________________

Choose a weather picture to stick into the space below. Then write some questions that you would like answered about this type of weather.

I would like to learn the following about ....

....

- ....

....

- ....

....

- ....

....

You can now use your questions to do some research about your chosen weather type. It may be useful to look in books and on the Internet.

How do we know what the weather will be like?

(This section links with pages 6 to 7 of the pupil book.)

Outcome

During this lesson the children will use a number of websites to investigate weather forecasting. They will then design their own weather forecast.

QCA Schemes of Work objectives
Unit 7, Section 6

Children should learn:

- to ask and respond to geographical questions
- to use geographical vocabulary about weather conditions around the world.

Key words and concepts

weather	meteorologist	temperature	humidity
air pressure	data	weather station	satellite

Lesson plan

1 Have the following resources available: computers with Internet access; individual photocopies of the map of the British Isles on page 30 of this teacher's guide.
2 Explain the lesson objectives to the children.
3 Ask the children to suggest why people want or need to forecast the weather and where the weather forecasters get their information.
4 Read pages 6 and 7 of the pupil book together.
5 Discuss any unfamiliar vocabulary and introduce the idea of weather maps and satellite pictures.
6 Ask the children if they or their families ever watch, read or listen to a weather forecast. Why might they want to know about the next day's weather? Are weather forecasts accurate?
7 Suggest some websites the children can visit to investigate weather forecasting. Good ones include www.bbc.co.uk/weather and www.met-office.gov.uk.
You could ask the children to find out:
   - What symbols are used for different types of weather in weather forecasting
   - What the weather is like at the moment
   - What the weather will be like for the next five days in your area.
8 Ask the children to complete the core activity (Worksheet 2).
9 Enable the children to think about evaluating the websites and the criteria they would use. They could use this evaluation to consider their own efforts.

Core activity – Worksheet 2
(page 7)

In this activity the children design their own weather forecast for the British Isles, using the websites they have been shown. The map of the British Isles on page 30 may be useful in this activity or the children could use the worksheet on page 7.

Follow-up activities

- Ask the children to collect local daily weather reports from the newspaper, TV, and Internet to help develop their understanding.
- Ask the children to find similar data for places in other countries and compare them with the forecast for a location in the UK.
- Ask the children to research an invention used to predict the weather.
- Encourage the children to write a jingle to accompany the weather forecast.

Exploring further

The children can find out more about forecasting the weather by using the following links on the Heinemann Explore Weather CD-ROM or website:

- Exploring > Weather outside your window > Observing the weather
- Digging deeper > Weather > A weather forecast
- Digging deeper > Forecasting the weather > Weather
- Digging deeper > Forecasting the weather > Ideas for forecasting weather
- Resources > Weather outside your window > Weather forecast

Worksheet 2

How do we know what the weather will be like?

Name ______________________________

Use the websites you have looked at to help you design a weather forecast for your local area or for the whole of the UK. Your forecast can either be for a day later this week or for a day in another season. Make sure your weather map and symbols are clearly presented. Use the space below to present your forecast.

How can we measure the weather at school?

(This section links with pages 8 to 9 of the pupil book.)

Outcome

During this lesson the children will undertake a fieldwork activity to develop their knowledge and understanding of the weather in their area. They will learn to use weather instruments and to take recordings accurately.

QCA Schemes of Work objectives
Unit 7, Section 4

Children should learn:

- to investigate places
- to ask and respond to geographical questions
- to use ICT to access information.

Key words and concepts

weather station **thermometer** **anemometer** **rain gauge** **microclimate**

Lesson plan

1 Have the following resources available: compass, (digital) camera, thermometer, equipment to make a rain gauge (outlined in the pupil book on page 8).
2 Explain the lesson objectives to the children.
3 Explain how to use a compass to find the direction of the wind.
4 As a class, plan a number of places around the school where the children will record the temperature, rainfall, and possibly wind direction.
5 Get the children to complete the core activity (Worksheet 3).
6 Get the children to complete the 'See for yourself' activity on pages 8 and 9 of the pupil book, i.e. actually measure the weather.
7 Ask the children to make predictions using their knowledge of the school grounds. For example, if they decide to record the temperature in a shady spot it is likely to be lower than in an open space.
8 Using the data collected, ask the children to explain any variations in rainfall, temperature, and wind direction and the effects of buildings, shade, etc. on their results.

Core activity – Worksheet 3
(page 9)

In this activity the children will plan to undertake a practical fieldwork exercise in the school grounds to measure temperature, rainfall, and possibly wind direction. As a class they will decide where they are going to record the weather. They will then all draw a map of the school grounds and make notes about each chosen site (e.g. shady site next to a wall, on the edge/in the middle of the playground, etc.) on Worksheet 3. They could use a (digital) camera to photograph each site.

Follow-up activities

- Make a class wall display using a large map of the sites around the school. You could include photographs, labels, and the data recorded.
- Ask the children to decide on the best site for a seated area/flower tubs or other items using the data collected. What would they need to do and how would they plan their investigation?
- Ask the children to write an instruction manual about how to make weather observations that will provide accurate data.

Exploring further

The children can find out more about measuring the weather by using the following links on the Heinemann Explore Weather CD-ROM or website:

- Exploring > Weather outside your window > Measuring the weather at school
- Digging deeper > Forecasting the weather > Recording weather data

Worksheet 3

How can we measure the weather at school?

Name ______________________________

Draw a map of your school grounds in the space below. Number the places where you are going to record the weather.

Write your notes about site conditions in the boxes below. For example, whether it is shady, in a corner, in the middle of the playground, etc.

Site 1

....
....
....
....
....

Site 2

....
....
....
....
....

Site 3

....
....
....
....
....

Site 4

....
....
....
....
....

Site 5

....
....
....
....
....

Site 6

....
....
....
....
....

Sunny weather

(This section links with pages 10 to 11 of the pupil book.)

Outcome

During this lesson the children will draw their own maps of places they like to visit in sunny weather after studying Ordnance Survey maps and aerial photographs.

QCA Schemes of Work objectives
Unit 7, Section 4

Children should learn:

- about weather conditions around the world
- to investigate places
- to ask and respond to geographical questions
- to use ICT to access information
- to identify similarities and differences

Key words and concepts

colour atmosphere deserts

Lesson plan

1 Have the following resources available: Ordnance Survey maps of your local area; aerial photographs of your local area; a large map of the local area.
2 Explain the lesson objectives to the children.
3 Ask the children to brainstorm with a partner the places they like to visit in their local area when it is sunny. Get them to think about why they like these places when the sun shines. Use this information in a class discussion.
4 Ask the children to locate their chosen places on a map and aerial photograph of the local area.
5 Ask the children to complete the core activity (Worksheet 4).
6 With the information gathered by the children, plot their favourite places on a large map.
7 Ask the children to use this to identify a favourite local place that the class could visit in sunny weather.

Core activity – Worksheet 4
(page 11)

In this activity the children will use Worksheet 4 to draw their own map of the local area and plot the places they enjoy visiting in sunny weather. They will use local Ordnance Survey maps to help them.

Follow-up activities

- Give the children the opportunity to use the Internet to find places outside the UK that have long periods of sunshine. Ask them each to choose one place and write a short description of the place and its weather.

- Ask the children to devise a questionnaire to ask classmates what place in the UK they would like to visit for a week if it was hot and sunny. Their questionnaire should give the opportunity to obtain reasons for their choices. Get the children to use this information to plot the destinations on a copy of the map on page 30 of this teacher's guide, and ask them to write a short report to present to the class. For a homework activity, suggest that the children use their questionnaire to survey their family members.

- On a sunny day ask the children to look carefully at the sky and record their observations by making notes, sketching, or taking photographs.

Exploring further

The children can find out more about sunny weather by using the following links on the Heinemann Explore Weather CD-ROM or website:

- Exploring > Weather outside your window > Sunshine
- Digging deeper > Sunshine and clouds
- Resources > Weather outside your window > Map showing the average hours of sunshine in the UK per day

Sunny weather

Name ______________________________

Draw your own map of the local area. Mark on your map the places you enjoy visiting in sunny weather. Remember to add a title and a key.

Rain and snow

(This section links with pages 12 to 13 of the pupil book.)

Outcome

During this lesson the children will write a poem about the rain or snow using some of the words they have generated in a brainstorming session.

QCA Schemes of Work objectives
Unit 7, Section 6

Children should learn:

- to ask and respond to geographical questions
- to use geographical vocabulary
- about weather conditions around the world.

Key words and concepts

precipitation **water vapour** **sleet** **hail** **drizzle**

Lesson plan

1. Explain the lesson objectives to the children.
2. Read pages 12 and 13 of the pupil book together.
3. Ask the children to brainstorm with a partner and list all the words that describe snow or rain (thinking about onomatopoeia and alliteration, etc.).
4. Discuss these words with the children before they begin the core activity.
5. Ask the children to do the core activity (Worksheet 5).
6. Ask the children to read out/present their poems and get the class to share in the evaluation of them.

Core activity – Worksheet 5
(page 13)

In this activity the children will use some of the words they have generated in the brainstorming session to write and illustrate a poem about rain or snow using Worksheet 5.

Follow-up activities

- The children could use oil pastels to produce rain/snow pictures for display with their poems.
- Ask the children to suggest ten different uses for snow.
- Get the children to name ten things they should not do in a blizzard.

Exploring further

The children can find out more about rain and snow by using the following links on the Heinemann Explore Weather CD-ROM or website:

- Exploring > Weather outside your window > Rain and snow
- Digging deeper > Rain and snow
- Digging deeper > Weather > Snow and ice
- Resources > Weather outside your window > An animation showing drizzle, rain, sleet, and snow
- Resources > Weather outside your window > Map showing average amount of yearly snowfall in the UK

Rain and snow

Name ______________________________

Use the words you have thought of to write a poem about rain or snow. Write your poem in the space below and illustrate it.

Windy weather

(This section links with pages 14 to 15 of the pupil book.)

Outcome

During this lesson the children will use secondary sources to help them locate places on a map. They will also understand the concept of prevailing winds.

QCA Schemes of Work objectives
Unit 7, Section 4

Children should learn:

- to investigate places
- to ask and respond to geographical questions
- to investigate similarities and differences.

Key words and concepts

air pressure | Beaufort scale | hurricane | prevailing wind

Lesson plan

1 Have the following resources available: atlases; board compass; compasses for the children.
2 Explain the lesson objectives to the children.
3 Read pages 14 and 15 in the pupil book together.
4 Ask the children to find their location on the map of the UK.
5 Use a large board compass to demonstrate to the children the eight compass points on the UK map and show the direction of the prevailing winds. This information can be found in an atlas or on the Heinemann Explore Weather CD-ROM or website (Resources > Weather outside your window > Map showing wind direction).
6 Get the children to use their compasses to find these directions in the classroom.
7 Ask the children to do the core activity (Worksheet 6).
8 Get the children to evaluate their work and suggest improvements.

Core activity – Worksheet 6
(page 15)

In this activity the children will use atlases to help them mark their home city/town/village on a map of the British Isles. They then indicate the direction of the prevailing winds. They will also need to devise a key and write a title.

Follow-up activities

- Ask the children to investigate what sort of weather we have if there is a prevailing east wind in summer. Suggest that they find out whether it would be the same or different from weather in winter and why.
- Encourage the children to use their compasses and what they learnt in the lesson to suggest the best place to fly a kite in the school grounds and locate this on a plan. Can they give reasons for their choice? Ask them to test and check their ideas on a windy day.
- Using the information on page 15 of the pupil book ask the children to find examples on the Internet or in news items of hurricanes or other evidence of the effects of the wind. This could be a homework activity.
- Get the children to list some ways to live safely in an area prone to hurricane activity.

Exploring further

The children can find out more about windy weather by using the following links on the Heinemann Explore Weather CD-ROM or website:

- Exploring > Weather outside your window > Wind
- Digging deeper > Wind and rain
- Resources > Weather outside your window > Map showing wind direction
- Resources > Weather outside your window > How winds are created

Windy weather

Name ______________________________

Use an atlas to help you locate and mark the place where you live on the map below.
Draw the prevailing winds on the map. Remember to add a title and a key.

Weather around the world

(This section links with pages 16 to 17 of the pupil book.)

Outcome

During this lesson the children will use secondary sources and the Internet to locate and investigate world climate zones.

QCA Schemes of Work objectives
Unit 7, Section 2

Children should learn:

- to ask and respond to geographical questions
- to use geographical vocabulary
- about weather conditions around the world.

Key words and concepts

equator	northern hemisphere	southern hemisphere	North Pole
South Pole	Tropic of Cancer	Tropic of Capricorn	Arctic Circle
Antarctic Circle	zone	temperate	

Lesson plan

1 Have the following resources available: atlases; reference books; Internet access; globe; torch.
2 Explain the lesson objectives to the children.
3 Explain to the children the difference between weather and climate.
4 Ask the children to give examples of the weather today and several contrasting days, or examples of how they have experienced different types of weather on a single day.
5 Using a globe, tell the children how the Sun's rays are more focused at the equator than in the polar areas. Show this with the torch and globe or by drawing (e.g. on the blackboard) two parallel bands of light reaching the equator and one of the polar regions.
6 Read pages 16 and 17 in the pupil book together.
7 Examine how the children will locate hot (tropical), cold (polar), and temperate zones.
8 Ask the children to do the core activity (Worksheet 7).
9 Invite the children to share their findings with the rest of the class.

Core activity – Worksheet 7
(page 17)

In this activity the children need to use an atlas to locate some countries near the equator. They should mark these places on the maps on Worksheet 7. They will use atlases, books, and the Internet to find out about the climate in each place. Ask them to locate and investigate some areas that are the coldest places in the world. They should mark these places on their maps. Get them to find information about the climate in these cold places.

Follow-up activities

- Ask the children to locate the UK on the globe and investigate the climate. Ask the children if they can locate a country with a similar climate to the UK.
- Suggest that the children choose one of the places located near the equator and another place in a polar region. Ask them to record five ways that the places are different. Are they able to suggest any similarities between the places?
- Ask the children to plan and trace a route on the globe of a journey around the world that starts and ends in the UK. Get them to record the continents and climate zones through which they pass.

Exploring further

The children can find out more about weather around the world by using the following links on the Heinemann Explore Weather CD-ROM or website.

- Exploring > Weather around the world > Desert weather
- Exploring > Weather around the world > Tropical weather
- Exploring > Weather around the world > Temperate weather
- Exploring > Weather around the world > Weather in polar regions
- Digging deeper > Weather > Hot and cold
- Resources > Weather outside your window > Map showing climates of the world
- Resources > Weather outside your window > Temperatures

Weather around the world

Name ____

- **Look in an atlas and locate three of the hottest countries in the world. Mark them on the map below.**
- **Now find two of the coldest places in the world. Mark these places on the map below.**
- **In the boxes at the bottom of the page write a short description of what the climate is like in one hot place and one cold place.**

The climate in .... is ....

....

The climate in .... is ....

....

How does the weather affect human activity?

(This section links with pages 18 to 19 of the pupil book.)

Outcome

During this lesson the children will develop their ability to work in a team and improve their thinking skills and understanding of the weather by making decisions about human activity.

QCA Schemes of Work objectives
Unit 7, Section 6

Children should learn:

- to ask and respond to geographical questions
- to use geographical vocabulary
- about weather conditions around the world.

Key words and concepts

dehydrated   flooding   suitable

Lesson plan

1 Explain the lesson objectives to the children.
2 Ask the children to read pages 18 and 19 of the pupil book.
3 Get the the children to work in groups to do the core activity (Worksheet 8).
4 On completion of the core activity the children should then work together in their group to prepare a presentation to make to the rest of the class.
5 Ask the children which of the headings they found most/least easy to write about and whether they can suggest why. Was this the same for all groups?

Core activity – Worksheet 8
(page 19)

In this activity the children will use Worksheet 8 to develop their thinking skills by deciding on activities that people are not able to do in each of the types of weather listed.

Follow-up activities

- Use the photograph on page 19 of the pupil book, or collect and use similar pictures, and ask the children to describe how they would feel if their home had been flooded. Ask them to think about the one thing they would save from their home if it were flooded. Ask them to give reasons for their choice.

- Ask the children to look at the photograph on page 19 of the pupil book and discuss what might have happened before the photograph was taken to cause what they can see and what might happen after the photograph was taken to help resolve the situation.

- Enable the children to create and perform a dance to represent a type of weather mentioned on pages 18 and 19 of the pupil book. They could develop their own musical accompaniment for this using available instruments or by designing and making their own.

Exploring further

The children can find out more about how the weather affects human activity by using the following links on the Heinemann Explore Weather CD-ROM or website:

- Explore > Wild weather > Flooding
- Digging deeper > Wild weather > Heatwave
- Resources > Wild weather > A video showing a blizzard
- Resources > Wild weather > A video of a tornado

Worksheet 8

How does weather affect human activity?

Name ____________________

Decide on some activities that people are *not* able to do in each of the types of weather listed below.

Hot weather

....

....

....

....

....

....

....

....

Rainy weather

....

....

....

....

....

....

....

....

Cool weather

....

....

....

....

....

....

....

....

Snowy weather

....

....

....

....

....

....

....

....

Why do people go on holiday?

(This section links with pages 20 to 21 of the pupil book.)

Outcome

During this lesson the children will work together to research a number of places, eventually deciding where they would most like to visit. They will use a variety of secondary sources to do this. They must also justify their conclusions.

QCA Schemes of Work objectives
Unit 7, Section 6

Children should learn:
- about weather conditions around the world
- to ask and respond to geographical questions
- to use geographical vocabulary.

Key words and concepts

weather, climate, destination, hottest, coldest, reasons, important factor

Lesson plan

1. Have the following resources available: newspapers/magazines that include articles on locations around the world.
2. Explain the lesson objectives to the children.
3. Read page 20 of the pupil book with the children.
4. Discuss with the children why people go on holiday, identifying that the weather is usually an important factor.
5. Give the children the newspaper/magazine articles and ask them to do the core activity (Worksheet 9) with a partner/in a group.
6. Discuss with the children the types of weather/climate in the different locations they have identified. Which were the hottest and coldest destinations they listed?

Core activity – Worksheet 9
(page 21)

In this activity the children will work with a partner or in a small group. They should look at a selection of newspapers and magazines. The children should look for four different locations around the world that are mentioned in the newspapers or magazines. They need to locate these places on a map and to find out about the weather in each place. They will then decide which place they would like to visit on a holiday. They must also think about why they have chosen this place.

Follow-up activities

- Ask the children to watch the news on television and list five places that are in the news that day. Ask them to note if the weather was mentioned in the news item. Find out about the weather in these places and locate them on the globe/map.
- Ask the children to imagine themselves in one of the places they have investigated. How would they feel about the weather in that place? Would there be any differences in the weather if they travelled there at different times of the year? Which type of weather/time of year would they prefer and why?
- Suggest that the children collate information about temperature and rainfall in one of the places they investigated over a particular time period. Get them to represent this in a number of graphical ways, including on a computer.

Exploring further

The children can find out more about how the weather affects where we go on holiday by using the following links on the Heinemann Explore Weather CD-ROM or website.

- Contents > Exploring
- Exploring > Weather around the world > Two very different temperate climates
- Digging deeper > Weather > Mediterranean climate

Worksheet 9 **Why do people go on holiday?**

Name ______________________

1. On the map below, mark four of the places mentioned in the articles your teacher gave you.

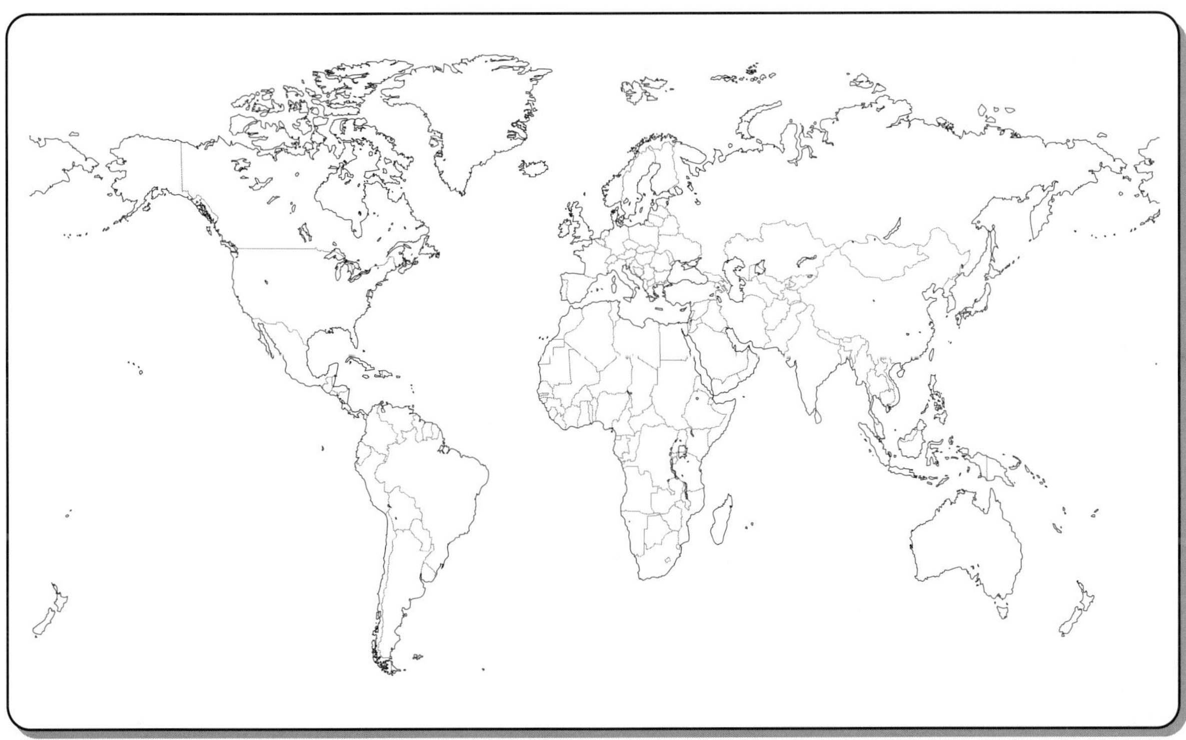

2. Write a few lines about the weather in each of these places.

Place 1

....
....
....
....

Place 2

....
....
....
....

Place 3

....
....
....
....

Place 4

....
....
....
....

3. Now decide which place you would most like to visit and say why.

....

Where can we go on holiday?

(This section links with pages 22 to 23 of the pupil book.)

Outcome

During this lesson the children will think about and discuss their own holiday choices before matching holiday destinations with a number of characters. They will make decisions using the information provided before discussing and evaluating their choices.

QCA Schemes of Work objectives
Unit 7, Section 3

Children should learn:

- about weather conditions around the world
- how places relate to each other
- to use and interpret atlases, maps, and globes
- to ask and respond to geographical questions.

Key words and concepts

safari   National Park   destination   activities   decision-making

Lesson plan

1 Have the following resources available: globes or blank photocopies of the worold map on page 31 of this teacher's guide.
2 Explain the lesson objectives to the children.
3 Read pages 22 and 23 of the pupil book together and talk about the holiday destinations mentioned.
4 Discuss the children's own holiday destinations and locate these on an appropriate map/globe.
5 Talk about how they travelled, how long the journey took, and if the weather affected their stay. Were any of their activities affected by the weather?
6 Ask the children to do the core activity (Worksheet 10). It may be helpful to enlarge the worksheet to A3 to allow room for the children to fill in the cards.
7 Ask the children to choose the person who was the most difficult/easiest to match with a destination. Can they suggest reasons for this?

Core activity – Worksheet 10
(page 23)

In this activity the children may work with a partner to complete Worksheet 10, which has a selection of holiday destinations and a number of characters. They need to match the characters with the destinations that might suit them best. They will need to make decisions using the information provided and talk about how they made their decisions. Finally the children will think about the details of each of the holidays mentioned.

Follow-up activities

- Ask the children to make up a list of alternative destinations and characters that would match them.
- Suggest that the children find some examples of music from one of the destinations discussed. Listen to it and try to identify the instruments being played.
- Ask the children to create and perform their own music using available instruments or by making their own.

Exploring further

The children can find out more about where we can go on holiday by using the following links on the Heinemann Explore Weather CD-ROM or website:

- Explore > Weather around the world > Tropical weather
- Digging deeper > Weather > The seasons

Where can we go on holiday?

Name ______________________________

Look at the holiday destinations on the luggage labels and match them to the people on the cards. You need to think about what sort of person would like to go there and what each of the people enjoys doing.

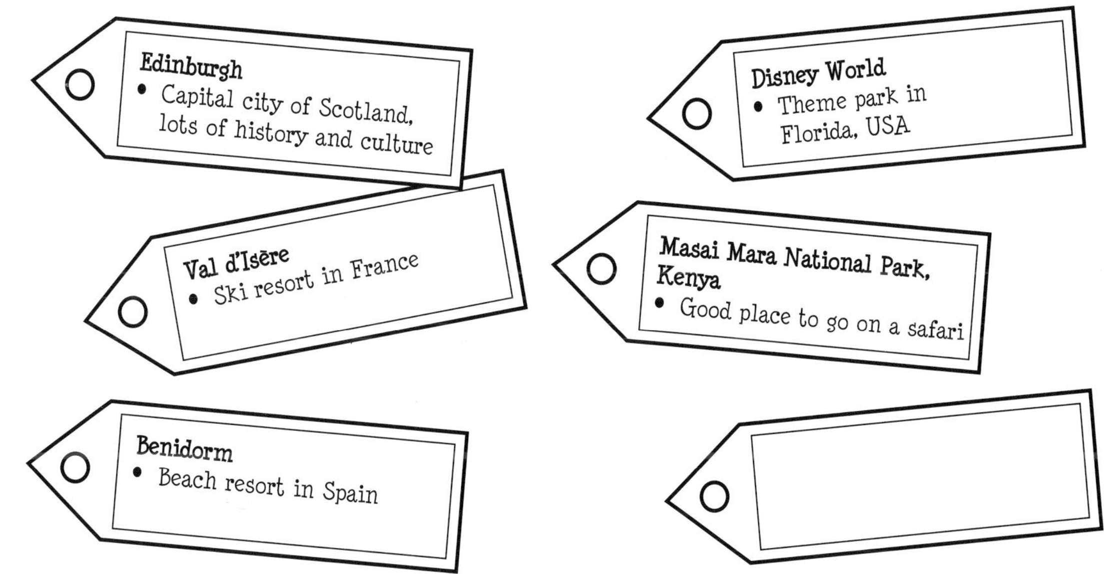

Choose a destination for the blank label and describe the person who would like to holiday there on the blank person card.

Janina	
Likes Animals Visiting the zoo	**Dislikes** Cold weather
Destination:	

Emily	
Likes Sports and being busy	**Dislikes** Beaches
Destination:	

Hamish	
Likes Museums and visiting the theatre	**Dislikes** Flying
Destination:	

Ben	
Likes Sunshine and swimming	**Dislikes** Museums
Destination:	

Ali	
Likes Fast rides and having fun	**Dislikes** Wild animals
Destination:	

Likes	**Dislikes**
Destination:	

Now think about:

- **How would they get there?**
- **When would they visit the destination?**
- **Where might they stay?**

What do we need to take with us?

(This section links with pages 24 to 25 of the pupil book.)

Outcome

During this lesson the children will discuss taking appropriate clothing on holiday according to the weather conditions.

QCA Schemes of Work objectives
Unit 7, Section 5

Children should learn:

- about the effect of the weather on human activity.

Key words and concepts

tropical polar temperature

Lesson plan

1. Have the following resource available: a globe/map of the world at the front of the classroom.
2. Explain the lesson objectives to the children.
3. Discuss local places of interest/attractions and what the children would wear/need to take with them at different times of year or in differing weather conditions (e.g. wet/sunny day in winter or wet/sunny day in summer).
4. Read and discuss page 25 of the pupil book.
5. Ask the children to spend a few moments carefully looking at the information.
6. Use the information to describe to the children what someone would be wearing on a visit to a place described on page 25 and ask them to quickly name and locate the correct place on the globe/map. Ask the children to suggest other countries where these clothes might also be suitable.
7. Ask the children to work in groups to complete the core activity (Worksheet 11).
8. Share some of the children's work with the rest of the class.

Core activity – Worksheet 11
(page 25)

In this activity the children will create their own chart similar to that on page 25 of the pupil book. They should choose different countries. Ask them to try to include examples from tropical, temperate, and polar regions.

Follow-up activities

- Ask the children to name ten things they would be unlikely to wear on a seaside holiday to Spain.
- Ask the children to list things they would not take to wear on a trip to Antarctica.
- Ask the children to compare and contrast what they usually wear and what is worn in one of their chosen localities.
- Suggest that the children create a collage to represent what they would wear when visiting the places they have researched.

Exploring further

The children can find out more about what we need to take with us by using the following links on the Heinemann Explore Weather CD-ROM or website:

- Explore > Weather around the world
- Digging deeper > Weather > Hot desert climates
- Digging deeper > Weather > The wet tropics

Worksheet 11

What do we need to take with us?

Name ______________________________

Complete the table below for some contrasting locations around the world. Try to include examples from tropical, temperate, and polar regions.
Make sure you use different locations to those listed in the pupil book!

Country	Continent	Climate	Clothes

Choose one of the locations above and draw a person visiting that place who is wearing appropriate clothing.

How are places similar to and different from our own locality?

(This section links with pages 26 to 27 of the pupil book.)

Outcome

During this lesson the children will contribute to the making of a large class chart, before working in groups to generate their own questions about places in the world.

QCA Schemes of Work objectives
Unit 7, Section 4

Children should learn:

- to investigate places
- to ask and respond to geographical questions
- to use secondary sources
- about weather conditions around the world
- to identify similarities and differences
- to use ICT to access information.

Key words and concepts

glaciers beautiful botanical

Lesson plan

1. Have the following resources available: atlases; globes; access to the Internet; books about different countries; old newspapers/magazines with articles about overseas destinations.
2. Explain the lesson objectives to the children.
3. Read pages 26 and 27 from the pupil book together.
4. Ask the children to use the information gathered in the 'See for yourself 'activity to fill in a large class chart for the local area and for the places mentioned on pages 26 and 27 of the pupil book.
5. Ask the children to do the core activity (Worksheet 12).
6. Give the children the opportunity to evaluate their work.
7. Enable the children to pool their work by producing a brochure as a class/school resource using ICT. Send this as an email attachment to a contact school.

Core activity – Worksheet 12
(page 27)

In this activity the children should choose a new place to investigate. They should answer the same questions that are given in the 'See for yourself' activity on page 26 of the pupil book. To do this they will need access to books, magazines/newspapers, and the Internet. They will then think about the similarities and differences between their own locality and their chosen destination and complete Worksheet 12.

Follow-up activities

- Ask the children to choose one of the places they have found out about and write a short illustrated article for a newspaper's travel/holiday supplement.

- Ask the children to create an advertisement for one of the places they investigated to get people to visit it.

- Ask the children to write a letter to a child in one of their chosen destinations explaining about their location, weather, and lifestyle.

Exploring further

The children can find out more about how places are similar or different from their own locality by using the following links on the Heinemann Explore Weather CD-ROM or website:

- Explore > Weather around the world > Two very different temperate climates
- Digging deeper > Weather > The British climate
- Digging deeper > Weather > Mediterranean climate

Worksheet 12

How are places similar to and different from our own locality?

Name ____

Use the globe or atlas to choose a place to investigate. Now try to answer the questions below.

- What is the name of the place? ....

- Where is it? ....
....

- What is the weather like? ....
....
....
....
....

- What is the landscape like? ....
....
....
....

- Are there any special features? ....
....
....
....

As part of your investigation you could find pictures that help to answer your questions. Now create a report, using any pictures you have found, comparing your chosen destination with your own locality. You will also need to collect or draw pictures of your own locality. Write your report on a separate piece of paper.

Which places have we visited?

(This section links with pages 28 to 29 of the pupil book.)

Outcome

During this lesson the children will plan and make a fact file for a place they have visited, or would like to visit.

QCA Schemes of Work objectives
Unit 7, Section 7, 4

Children should learn:
- to investigate places
- to use and interpret atlases and maps
- to use ICT to access and present information.

Key words and concepts

continent weather climate desert temperate
Arctic Circle fact file

Lesson plan

1 Have the following resources available: atlases; maps; reference books; access to the Internet.
2 Explain the lesson objectives to the children.
3 Read pages 28 and 29 in the pupil book together and discuss making a fact file.
4 Ask the children to do the core activity (Worksheet 13).
5 Get the children to evaluate the work in progress to assess whether they wish to make any changes to their fact file.
6 On completion, display the finished fact files around a world map.

Core activity – Worksheet 13
(page 29)

In this activity the children will make a fact file about a place they have visited or would like to visit. They will include information on place, country, continent, climate zone, and other factual data, such as population. They need to focus on accuracy and clear presentation. Atlases, maps, books, and the Internet can be used to help them research all the information they need. They could use ICT to present their work.

Follow-up activities

- If they have used ICT to preent their work, the children could send their completed fact file as an email attachment to another school or a friend in another class in their own school.

- If the children have produced their completed fact files using ICT, an extra set could be printed to use for sorting and matching activities, particularly if they were laminated. The cards could be sorted into similar climate zones or the children could devise their own criteria.

- Ask the children to devise and write a set of instructions for a game using their set of fact files. Get them to swap their sets of instructions with a friend, play and evaluate each other's games.

Exploring further

The children can find out more about some of the places they have investigated by using the following links on the Heinemann Explore Weather CD-ROM or website.

- Explore > Weather around the world
- Digging Deeper > Weather

Which places have we visited?

Name ______________________________

Make a fact file about a place you have visited or would like to visit. You may need to do some research to find out all of the information you need.

- Official name: ....
- Capital city: ....
- Population: ....
- Continent: ....
- Climatic zone: ....
- Official language: ....

Flag

Map

Weather

Clothes I might wear on my visit

If you would like to include photographs you might like to create a bigger fact file on a separate piece of paper.

Blank map of the British Isles

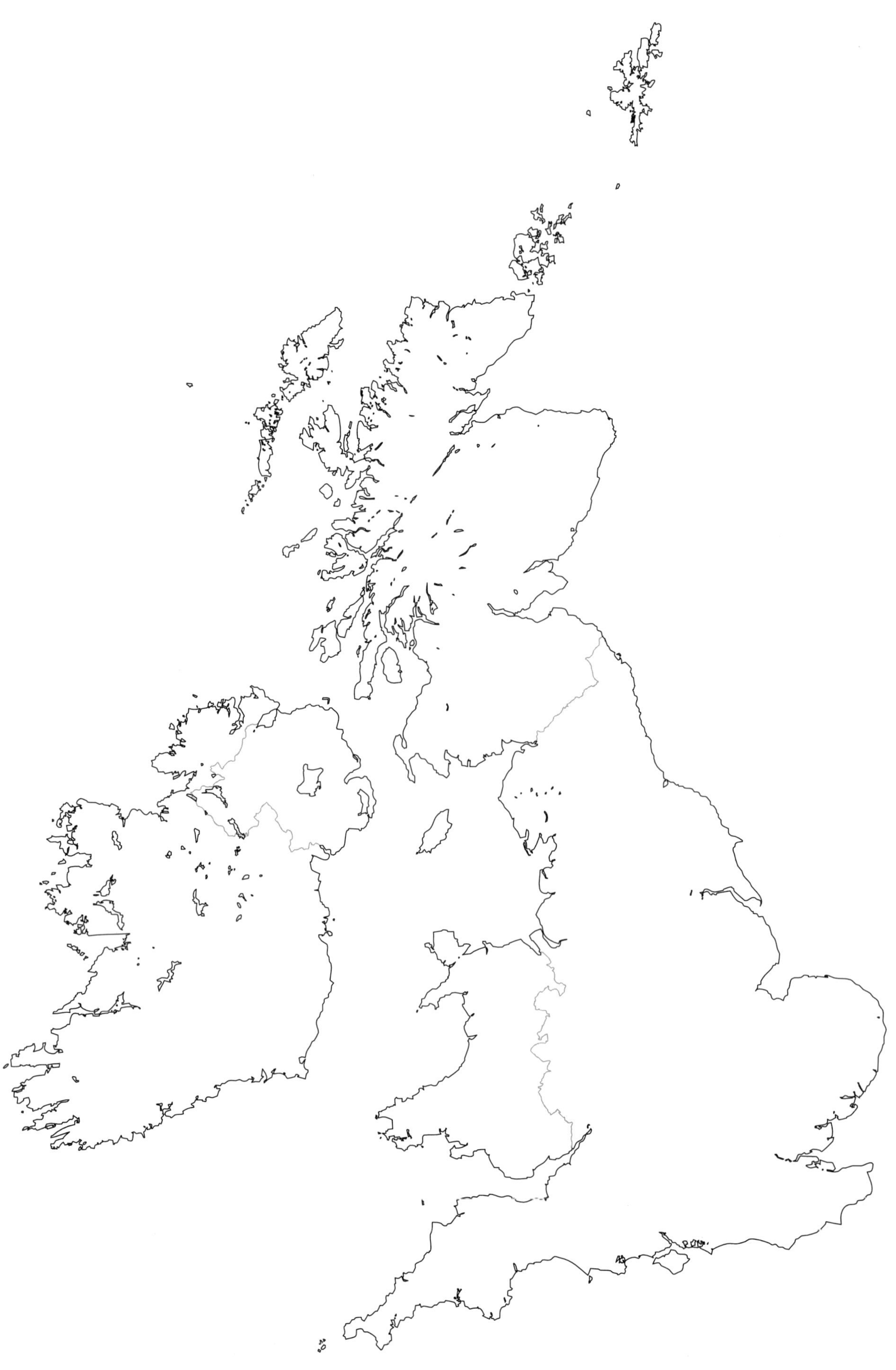

Blank map of the world

Further research

Here we have tried to provide a range of relevant resources for both pupils and teachers. The list is not intended to cover everything available, but to give ideas about resources that will complement the material in Explore Geography.

Books for pupils

There are many children's books available on the weather, from non-fiction books for information and research, to stories and poetry.

Discovering Geography: Weather, Rebecca Hunter (Raintree, 2004)
This is one of a series of books on various geographical topics. It provides an engaging introduction to the weather. The book includes four easy and intriguing projects that illustrate important principles and spark discovery-based learning.

e.explore: Earth (Dorling Kindersley, 2004)
A book with an accompanying website that has live updated information and images to download, plus access to animations, videos, and sound buttons.

Awesome Forces of Nature: Howling Hurricanes and *Terrifying Tornadoes*, by Richard and Louise Spilsbury (Heinemann, 2003)
These books look at the causes of hurricanes and tornadoes, how professionals use weather data to predict them, and the impact they have on people's lives.

A Year Full of Poems (Oxford, 1996)
This book contains many poems about the weather and seasons by various poets and has colourful illustrations.

Letters Around the World, Thando MacLaren (Tango Books, 2004)
A book about children in different places around the world writing as pen pals.

A Balloon for Grandad, Nigel Gray (Orchard Books, 2002)
A story about a windy day and a balloon's journey from a little boy's home in the UK to his grandfather's home far away in Africa.

Take your camera to ..., series by Ted Park (Raintree, 2003)
A young person's travel guide to some of the world's most interesting countries.

Resources for teachers

The QCA Schemes of Work for Geography and guidelines for using them can be found online at www.standards.dfee.gov.uk/schemes.

The Primary Geography Handbook (Geographical Association, 2005)
This book has many practical ideas for teaching primary Geography. The aim of the book is to inspire teachers who are both geography experts, and those who are new to the subject.

Guide to the Weather, Ross Reynolds, (Philip's, 2002)
A practical guide to observing, measuring, and understanding the weather. It can be used as an aid to planning holidays with a country-by-country climate guide.

Websites

wwww.bbc.co.uk/schools/barnabybear
This site has a simple weather report game included in a wide range of geography activities.

www.bbc.co.uk/weather/
An excellent site giving access to weather information on a local, UK, and worldwide basis.

www.onlineweather.com/v4/world/owac/index.html
This site has local, UK, and worldwide weather forecasts clearly presented. Detailed data is available for comparison with the children's own work.

www.incnetwork.demon.co.uk/factfile/weather.htm
This site lists and gives access to good UK weather websites.

www.met-office.gov.uk
Good UK weather site with current satellite images, data, etc.

www.geography.org.uk
Current ideas, information, and resources are available on this site.

www.sin.org.uk/geography
This site gives access to ideas and information for teachers and children.